Mars

by Patrick H. Stakem

(c) 2018

Number 19 in the Space Series

Table of Contents

Introduction..4
Author..4
A note on Units...5
Schiaparelli and the Canals..6
Mars' moons...6
Von Braun's and Ley's vision 7
Getting to Mars...8
The Mars Environment...9
Mars Missions...12
   United States Missions to Mars...12
      Viking..13
      Mars Pathfinder...15
      Sojourner..16
      Mars Climate Orbiter...16
      Mars Polar Lander...18
      Mars Odyssey...18
      MER – Mars Exploration Rovers Spirit & Opportunity 18
      Mars Reconnaissance Orbiter..20
      Phoenix Lander...21
      Dawn...22
      Mars Helicopter Scout...22
      Mars Science Laboratory Curiosity....................................24
      Maven...27
      Insight...28
      Interplanetary Cubesat..28
   The Russian/Soviet Mars Missions..30
      Phobos-Grunt..31
   ESA Missions..33
      Mars Express..33
      Rosetta...33
   The Indian Mars Mission...33
   United Arab Emirates..34
   Japan..35
   Chinese Mars Mission 35
Future Missions...36

- Exo-MARSrosalind...........36
- Mars-2020...........37
  - Perseverance and Ingenuity...........37
  - Planned Crewed Mars Missions...........38
    - Space-X...........38
    - NASA...........39
- Getting to Mars...........40
  - Getting around Mars...........41
  - Interplanetary Internet...........43
- Crewed Missions...........45
- Human Missions to Mars...........46
  - Orion Capsule...........47
  - Deep Space Gateway...........47
- Infrastructure...........48
- Colonization of Mars...........49
  - Mars Analog Bases on Earth...........49
  - Mars Base Camp...........50
  - Mars Crewed Fly-by...........54
- Afterword...........55
- Bibliogaphy...........55
- Resources...........63
- Glossary of terms and definitions...........69
- If you enjoyed this book, you might also be interested in some of these...........78

Introduction

Missions to Mars and beyond are lengthy, expensive, and dangerous. The Red planet has fascinated mankind for thousands of years. Many thousands of years of position data have been recorded. We have robotic spacecraft in Martian orbit, and landers and rovers on the surface. The next step is to go there, with boots on the ground. Decades of time, and hundreds of millions of dollars are required. Columbus would have approved. We just need to get off our...planet, and get going.

At this writing, every one is focusing on the Artemis Lunar Missions. Getting boots on the lunar regolith is a short term goal. We're learn a lot, and the technology's we develop can be used to get to Mars (and back safely, or decide to stay).

Author

The author's Aerospace career has revolved around support for space-based microprocessors and computers for NASA since 1971.

Mr. Stakem received a Bachelor's Degree in Electrical Engineering from Carnegie Mellon University, and masters in Physics and Computer Science from the Johns Hopkins University. He has followed a career as a NASA support contractor, working at every NASA Site. He taught for the Graduate Computer Science Department at Loyola University in Maryland, the Whiting School of

Engineering of the Johns Hopkins University, and Capital Institute of Technology.

I had the very good luck to meet and talk with Willy Ley. Curiously, I worked for von Braun. We were at Fairchild Aerospace at the same time, but I never go to meet him before he died.

Mr. Stakem can be found on Facebook and Linkedin.

A note on Units

I am fairly conversant in both English and Metric units (what is the metric equivalent of furlongs per fortnight?). Metric (SI) is mandated for NASA usage now, for interchangeability with our partner space faring nations. When a lot of the legacy flights discussed here were flown, English units were the norm. I have tried to keep the units comparable to the mission at the time. Conversions are easy enough, but units conversion is a source of error. It has caused loss of missions and lives. It's not what you know about units and measurement, its how you think. And, I still think English units (even the English use Metric now), and convert in my head or on my phone.

For scientific/engineering work, the Metric system is well thought out. For artisans, the English system served well, as most units were divisible by 2. Which is easy. Fold the cloth. Hopefully, when we are all taught Metric first, some one will still remember the conversions. You just need a good slide rule....

Schiaparelli and the Canals

The story of Mars is not complete without a discussion of Schiaparelli's "discovery" of the Martian canals. He was an Italian astronomer, from the University of Turin, who spent 40 years at an observatory in Milan. In 1877, he observed and reported a network of linear structures on the planet, and termed them "canali", the Italian word for channels. He wrote an 1893 book, "Life on Mars." A strong supporter of life on Mars was Percival Lowell, at the observatory in Flagstaff, Arizona. The later Italian astronomer Cerulli decided the channels were optical illusions, This seems to have been born out, due to the many close up observations of the planet by spacecraft. Schiaparelli went on to work on binary stars, and the linkage between comets and meteor showers. He did extensive work on the observation of Mars and Venus., and kept careful notes and drawings.

Mars' moons

Mars has two moons, Phobos and Deimos. These were discovered in 1877 at the U. S. Naval Observatory in Washington, D. C. by Asaph Hall. He was using the observatory's 26 inch refracting telescope. The names come from the son's of the Greek god of War, Ares. Two moons of Mars are mentioned by Jonathan Swift in Gulliver's Travels, and there was wide speculation of their existence. Many of the surface features on the moons are named after characters in Swift's Sci-Fi novel.

Phobos orbits Mars with a semi-major axis of 9375 km and an orbital period of 7.65 hours. Deimos has a semi-major axis of 23,450 km and orbital period of 30.35 hours. Viewed from Mars' surface, Phobos appears about 1/3 the size of our full moon. Due to distance and geometry, there are total lunar eclipses of Phobos almost every night. The small size of the moons means there are no total solar eclipses on Mars. Both moons are in tidal lock with Mars, so only one side can be seen from the surface. There is an interesting feature called the Phobos monolith. It is a rock about 85 meters across.

Phobos rises in the west and sets in the east, while Deimos does the opposite. Deimos's altitude is above synchronous, and takes it 2.7 days to fall below the horizon. The orbital period for Phobos is 7.6 hours, and 30.3 hours for Deimos.

One theory is that the moons are captured asteroids, another is that they resulted from a collision. Both have been extensively photographed, but no samples have been taken as of this writing.

Mars has a series of Trojans, which are objects that share it's orbit around the Sun. They are clustered around the leading (L4) and trailing (L5) Lagrange points, at 60 degrees ahead and behind the Planet. There are seven known Trojans. Since the Lagrange points are gravity nulls, stuff tends to accumulate there.

Von Braun's and Ley's vision

Wernher von Braun produced a detailed technical project plan for getting to Mars in 1952, but he had been thinking about it long before that. He and fellow rocket enthusiast Willy Ley worked on Mars Mission Concepts, a long time before the first person rode a rocket.

Getting to Mars

Getting to Mars is not easy, and requires a lot of energy and time. First, we get to Earth orbit, and check that all the systems have survived launch. Knowing where Mars is currently, we can calculate the correct point to head to, when our mission reaches the correct distance. When the mission reaches Mars, we have to slow down and enter Martian orbit. Counter-intuitively, a minimum energy mission (but not a minimum time scenario) takes us looping past Venus to get to Mars in an efficient manner.

The distance from Earth to Mars varies on their relative positions in their orbits, but ranges from 33.9 to 250 million miles.

Every 26 months, the Earth and Mars align such that a minimum energy transfer can get a spacecraft from one to the other. The implications are that if you miss the launch window, you need to wait some 2 years, and that there is a minimum time for a Earth to Mars journey, landing, and return to Earth. This sets a mission duration that is important for future crewed missions. The travel time for this path is about nine months.

A Hohmann transfer orbit is an elliptical orbit between the circular orbits of two planets. It allows for the maximum payload to be sent with a given energy expenditure (or, amount of rocket fuel). You can get there quicker by expending more energy. The alignment, or launch window, is critical. The details of the interplanetary transfer orbit were defined and published by German scientist Walter Hohmann in 1925.

The hope is, we can establish a refueling station on Mars or one of its moons, so we don't need to carry our return fuel with us. Since Mars has been proven to have water ice, this might be quite feasible. Using solar energy, you can break down water into liquid hydrogen and liquid oxygen – rocket fuel and oxydizer. NASA has published a map of water ice on Mars, and ESA's Mars Express spacecraft spotted the remains of an ancient river system and lake.

The Mars Environment

Mars is the next planet out from the Sun, from the Earth. It is almost the smallest, but Mercury has that title. The red of the red planet comes from iron oxides, the ore hematite. Having a slight gravity, Mars has a thin atmosphere. This means that many things evaporate into space, such as water. If water exists on Mars, it is in the form of ice. Indeed, the Martian poles have water ice. There was a large volume of underground ice discovered in the area of Utopia Planitia. Mars can be seen by the naked eye, from Earth's surface. Mars scores the largest the largest known volcano in the solar system, Olympus Mons. It sdticks up through the atmosphere. Mars' spin axis is tilted like the Earth's, so it has seasons. It consists mostly of desert terrain, with tall mountains and impact craters. No oceans, no canals. It has an almost non-existent magnetic field to ward off charged particles.

The one-way light time between Earth and Mars varies between 3 and 29 minutes, depending on the relative positions in their orbits. This affects the communication between the Earth, and the various orbiters, landers, and rovers. In some cases, when Mars and the Earth are on opposite sides of the Sun, communication is not possible. Optical telescopes on Earth can resolve features down to a size of 300 km, due to the distortions of Earth's atmosphere. This drives the many imaging missions that have visited the red planet.

Mars' size is about half of Earth's, but it has about the same dry-land surface area. It has a metallic core, but not a significant magnetic field. We have good data on the composition of the surface courtesy of the Phoenix and other landers. Surface features from a variety of sources seem to show erosion. More water certainly existed on Mars in earlier periods. The poles are very cold, and covered in solid carbon dioxide. The United States Geologic Survey maintains accurate maps of the surface. Mars has a large number of impact craters, due to the thin atmosphere.

Mars has polar ice caps, of frozen water ice, and sections of carbon dioxide. When the poles are in Sunlight, this "snow" or ice sublimates. This action results in clouds, and geysers of CO_2 gas. Mars gets less than half the amount of sunlight as does the Earth. When Mars is closest to the Sun, large dust clouds are generated, that can envelope the planet. One of these happened in 2018, affecting the operation of surface rovers. Mars has an observing weather satellite in orbit.

The Martian surface is blasted by the solar wind, due to its thin atmosphere, and lack of a magnetic field. Most of the Martian atmosphere is carbon dioxide. Methane is present, suggesting a biological origin, but other sources are possible. The current Indian Mars mission is specifically searching for methane in the atmosphere. Mars has awesome dust storms, some that cover the entire planet.

With Mars only receiving about 45% of the solar energy that the Earth gets, the surface can warm up during the day, but it always plummets at night, usually to where the carbon dioxide in the atmosphere freezes. Wind speeds can get up to 60 mph or so, but the atmosphere is so thin, you many not notice.

Mars was one of the items of interest for early astronomers. The Greeks and Romans named it after their god of war. The Sumerians called it Nergal, after their god of war. In Niveveh, the cult of the red planet called it "the star of judgment of the fate of the dead." The Egyptians were well aware of the planet, and the strange path it took across the heavens. Aristotle noted that Mars disappeared behind the moon, proving it was further away. Galileo was the first to study Mars with a telescope. The Chinese called it the fire star.

If you look in the Mayan document called the Dresden Codex, you will see an extensive list of observations of Mars, what we now call an Ephemeris. There is an observatory at Chichen Itza, in the Yucatan in Eastern Mexico.

Some asteroids, chunks of rock from Mars, are usually found in Antarctica. My old college college professor figured out why, and wrote the definitive paper. I haven't a clue.

One thing that would make Mars more habitable for humans has been very bad for Venus, and is a problem for Earth. Greenhouse gases on Mars would add density to the atmosphere, and control the wide swings of temperature. Not quite ready for terraforming.

Mars Missions

As of this writing, five Nations have sent missions to Mars: The United States, Russia, China, the Emirates, and India. The ESA, consisting of a number of European nation's, has also implemented Mars missions.

United States Missions to Mars

The United States has been involved in Mars exploration with missions starting in 1964, and continuing, as of this writing. That's more than 50 years of exploration, at a great distance. There have been 49 missions to the red planet.

The Mariner-3 mission to Mars in November of 1964 suffered a mission failure when the payload fairing failed to release. Mariner-4, launched 2 weeks later, was a success, as was Mariner-6 in 1969. (Mariner-5 went to Venus). Mariner 7 did a successful fly-by in 1969, but Mariner 8 had a launch failure.

Mariner-9 was launched in 1971, and entered Martian

orbit. It was a successful mission, and was deactivated 516 days later.

Viking

The Viking program was a pair of spacecraft sent to Mars in 1975. Each spacecraft consisted of an orbiter and a lander. The Viking landers used a Guidance, Control and Sequencing Computer (GCSC) consisting of two Honeywell 24-bit computers with 18K of memory, while the Viking orbiters used a Command Computer Subsystem (CCS) with two custom-designed 18-bit bit-serial processors. They were programmed in assembly language. The Honeywell machine had 47 instructions, and used two's complement representation for data.

After entering Mars orbit, the spacecraft extensively imaged the surface for a month. A target landing site was selected, and the landers went on their ways. The orbiters served as communication relays for the landers. They deployed solar panels for power, rated at 620 watts, at Mars' distance from the Sun. Viking-1 did 1385 orbits, while Viking-2 did 700. Viking-1 operated for 2245 Sols and Viking 2 lander achieved 128.

The Landers were stationary platforms, not rovers. They used parachutes and retro-rockets to safely reach the surface of the planet. For power, the lander platforms had radioisotope thermoelectric generators, providing 30 watts continuously. There were a variety of instruments including cameras, wind direction and velocity sensors, and a mass spectrometer. The instruments that targeted the detection of life activities did not see any significant results. The Viking 2 orbiter operated until July 1978, and its lander operated until April of 1980. The Viking 1 orbiter lasted until August of 1980, and the lander until November of 1982. Viking kicked off the search for organic molecules on Mars in 1976.

The landers are still on the surface, and could be moved to the Martian Museum of Exploration (when that is built).

The Mars Observer spacecraft suffered a launch failure in 1992. It was also termed the Geoscience/Climatology Orbiter. Three days prior to Mars orbit insertion, communication with the spacecraft was lost.

Mars Global Surveyor, launched in 1996, operated successfully for 7 years. It imaged the surface to uncover landing sites, and served as a communications relay back to Earth for surface assets. It also served as a Martian weather satellite, focusing on large dust storms that form on the surface. It had completed its primary mission in January of 2001, but continued to operate until November of 2006.

Surveyor was kept in a sun-synchronous orbit, where it passed over a given point on the surface at the same time (and Sun angle) each day. Surveyor was responsible for confirming the existence of water ice on Mars. It also photographed the Mars Exploration Rover *Spirit* and its wheel tracks on the surface.

Mars Pathfinder

The Mars Pathfinder mission landed on Mars on July 4, 1997. It carried a Rover named Sojourner, which was a 6-wheeled design, with a solar panel for power, but the batteries were not rechargeable. The rest of the lander served as a base station. Communication with the rover was lost in September. The Rover used a single 8-bit CPU with 64k of ram. It communicated with Earth via the base station using a 9600 baud UHF radio modem. The communication loss leading to end of mission was in the base station communication, while the Rover remained functional. The Rover had three cameras, and an x-ray spectrometer.

The computer in the mission base station on Mars was a single RS-6000 32-bit CPU of the IBM POWER architecture. The software was the VxWorks operating system, with application code written in the c language. The base station computer experienced a series of resets on the Martian surface, which lead to an interesting remote debugging scenario.

The operating system implemented pre-emptive priority thread (of execution) scheduling. The watchdog timer caught the failure of a task to run to completion, and caused the reset. This was a sequence of tasks not exercised during testing. The problem was debugged from Earth, and a correction uploaded.

The failure turned out to be a case of priority inversion. The higher priority task was blocked by a much lower priority task that was holding a shared resource. The lower priority task had acquired this resource and then been preempted by several medium priority tasks. When the higher priority task was activated, it detected that the lower priority task had not completed its execution.

The reset had the effect of wiping out most of the data that could show what was going on. This behavior was not seen during testing. It was successfully debugged and corrected remotely by the JPL team. Interplanetary cntrl-alt-delete.

Tasked with working for one month on the surface, it operated for three months, until September 1997.

Sojourner

The first wheeled vehicle on Mars was Sojourner, in 1997. It was also the first rover to operate on another planet. It was a 6-wheeled design, with a solar panel for power, but the batteries were not rechargeable. It lasted from July through September. It had a complex problem in its on-board computer, that was diagnosed and corrected from Earth. It also was expected to work for a month, but put in 3 months.

Mars Climate Orbiter

The spacecraft was lost on Mars in September 1999. The requirements did not specify units, so JPL used SI (metric) units and the contractor Lockheed Martin used English units. This was not caught in the review process, and led to the loss of the $125 million mission. The spacecraft crashed due to a navigation error. The computer architecture was a single 32-bit RAD6000 cpu, with 128 megabytes of ram, and 18 megabytes of flash memory. Check your units. In fact, if its science or engineering, Metric should be mandated.

VxWorks, from Wind River systems, was the operating system with flight software developed at Lockheed Martin.

Sensors and Actuators included dual 3-axis gyros, a star tracker, dual sun sensors, eight thrusters, and four reaction wheels.

The primary cause of the failure was human error. Specifically, the flight system software on the Mars Climate Orbiter was written to calculate thruster performance using the metric unit Newtons (N), while the ground crew was entering course correction and thruster data using the Imperial measure Pound-force (lbf). This error has since been known as the *metric mix-up* and has been carefully avoided in all missions since by NASA.

"The root cause of the loss of the spacecraft was the failed translation of English units into metric units in a segment of ground-based, navigation-related mission software, as NASA has previously announced," said Arthur Stephenson, chairman of the Mars Climate Orbiter Mission Failure Investigation Board. "The failure review board has identified other significant factors that allowed this error to be born, and then let it linger and propagate to the point where it resulted in a major error in our understanding of the spacecraft's path as it approached Mars."

Mars Polar Lander

The Mars Polar lander didn't do well in 1999. It took with it the Deep Space 2 probes. The hard landing may have been due to a premature engine cut-off, based on confusion between English and Metric units. Well, it was only a $125 million mistake.

Mars Odyssey

The Odyssey, launched in 2001, remains operational as of this writing. It is expected to remain so through 2025. It was deployed to search for surface ice, and serve as a communications relay. It has been for some time the main communication link between landers and rovers on the surface, and Earth. It is also tasked with characterizing the surface radiation environment for future crewed surface missions.

Mars has no magnetic field to speak of and a thin atmosphere. Both ionizing radiation from the Sun and charged particles hit the surface. During a solar storm, the levels can double.

It is in Sun-synchronous orbit around Mars. It helped to determine a landing site for the Mars Science Laboratory, and is expected to remain operational through 2025.

MER – Mars Exploration Rovers *Spirit & Opportunity*

The MER are six-wheeled, 400 pound solar-powered robots, launched in 2003 as part of NASA's ongoing Mars Exploration Program. *Opportunity* (MER-B) landed successfully at Meridiani Planum on Mars on January 25, 2004, three weeks after its twin *Spirit* (MER-A) had landed on the other side of the planet. Both used parachutes, a retro-rocket, and a large airbag to land successfully, after transitioning the thin atmosphere of Mars.

For power, they use 140 watt solar arrays and Li-ion batteries. The Rovers require 100 watts for driving, One problem that was noted was that the Martian dust storms cover the solar panels with fine dust, reducing their efficiency. This resulted in the use of a radioisotope generator on a subsequent mission. It's been observed that Rovers often use more energy in path planning, than to execute the actual path.

The onboard computer is a 20 MHz RAD-6000 32-bit CPU with 128 MB of RAM. There is a 3-axis inertial measurement unit, and nine cameras The Rovers communicate with Earth via a relay satellite in Mars orbit, Odyssey or Mars Global Surveyor. They also have the ability to communicate directly, at a lower data rate.

The lander and rover include Xilinx FPGA's for mission critical functions. The rover's motor controllers are implemented in the FPGA's. These operated through the worst Solar Flare ever measured. These are particularly bad at Mars, since it has no discernible magnetic field, or the equivalent of the Earth's Van Allen Belts. The FPGA's have SEU detection and use scrubbing for error correction. The Rovers have solar panels generating around 140 watts.

The Spirit unit became stuck in 2009, and engineers were unable to free it after 9 months of trying. It was re-tasked as a stationary sensor platform. Contact was lost in 2010. The Opportunity Rover lasted until February of 2019, after some 15 years of service

This mission was originally planned for 90 days, but the *Opportunity* Rover was still collecting useful data regarding potential life on our sister planet some 11 years later. It has traveled over 35 kilometers on the Martian surface. Ground based test units are used at JPL for evaluating problems seen on Mars, and for evaluating software and procedural fixes. Opportunity ceased operation in February 2019, having a fruitful mission of some 15 years.

In June of 2018, a massive Martian dust storm covered Curiosity, most importantly its solar panel, with a large amount of duct. The last communications was on June 10.

Mars Reconnaissance Orbiter

This mission launched in 2005, and remains operational at this writing. It was only scheduled to operate until 2008. By 2010, it had already transmitted more than 100 terabits of data to Earth. It joined five other active spacecraft at Mars. It serves as a telecommunications relay, and a weather mapper. It is searching for ideal landing sites for future crewd missions. It is modelled after the Mars Global surveyor.

MRO has two highly efficient solar panels for power, each panel generating a thousand watts. There are dual 50 ampre-hour batteries. The main computer is a rad-hard Rad-750, a special version of a PowerPC 750.

MRO has determined that Mars' north polar ice cap holds a total of 30% of the capacity of the current Greenland Ice Cap. MRO has also imaged the Phoenix lander on the surface, and the tracks of the Opportunity rover. MRO has confirmed the existance of flowing salty water on Mars.

Phoenix Lander

This was launched in 2007, and landed on Mars in 2008. Its mission ended in November of that year. It's mission was to search for microbial life, and to look for water. It completed its mission in August of 2008. After the Martian winter, the craft could not be contacted. It had a 90 day projected life on the surface, but exceeded this by 2 months. It used solar panels for power. It can relay data to Earth via the orbiting Mars Odyssey, MRO, and ESA's Mars Express.

Its landing was observed by these three spacecraft as well. The lander is a stationary platform, not a rover, and has a robotic arm and onboard analyzers. It dug a bit, and was able to confirm the existence of water ice just under the surface.

It had a three month defined mission, and exceeded this by 2 months. It succumbed to a dust storm, that covered its solar panels. Before this, it found snow falling from clouds.

Dawn

The Dawn Mission launched in 2007, and completed a Mars fly-by in 2009. It then continued onto to asteroids Vesta and Ceres in 2012. It has been orbiting Ceres since 2015. It had a software glitch in 2015, but the error was traced and worked around. It finished it's Ceres mission, and went into extended mission in 2016. to do a fly-by of asteroid 2 Pallas.

It used Mars for a gravity assist flyby. This stolen energy is estimated to have cost Mars a loss in velocity of 2.5 cm over 180 million years. Tough it out, Mars. The Dawn mission is still operating at this writing.It was to visit 145 Adeona in May 2019, but the Mission Senior Review Panel declined. It did get a one year extension at Ceres.

By June 2017, its working years were done. It went into an uncontrolled but stable orbit around Ceres. It ran out of hydrazine fuel in October 2018. It's still orbiting Ceres, its Mission done.

Mars Helicopter Scout

The next mission in 2021 included the 2020 Rover, which has an autonomous robotic helicopter named

Ingenuity. It is an eye-in-the-sky, looking out for hazards, planning a path, and seeing things that the rover's camera can't. It was to be a 10-day technology demonstration, planned to fly five times, during the early mission. It has completed its 19th flight. The copter blades are a meter in length and it has two coaxial counter-rotating sets. The range and flight duration have been exceeded. NASA says its mission is extended to September of 2022.

The atmosphere on Mars at ground level is about what it is on Earth at 87,000 feet.

The upper sensor assembly is on the mast, close to the center of mass. It has a commercial grade inertial measurement unit, and an inclinometer. The lower sensor assembly has a Garman Lidar,2 cameras, and a second IMU. The unit uses visual odometry, since we haven't gotten the GPS set up yet.

Compasses can't work on Mars due to the low magnetic field, so it uses solar tracking and inertial guidance for navigation. It has its own solar panels. It is carried under the rover, then dropped to the ground, and the rover moves some distance away so it can ascend.

It's computer, Qualcomm Snapdragon 901 (Arm architecture) runs linux. There are two microcontrollers. The battery's are Sony Lithium-ion 40 watt-hours. It has two identical radios, with SiFlex 02 chipsets, It can relay data implementing the Zigbee communications protocols.

It can relay data at 250 kilobits per second, over 1,000 meters. The Omni antenna is on the solar panel.

The helicopter was built by AeroVironment for JPL.

They have been designing and flying drones for some time.

NASA names the site where the helicopter first took off from as Wright Brothers Field. The U.N. agency International Civil Aviation rganiozation gave it an asirport code of JZRO (derived from Jezero Crater).

A successor mission is in the planning stage, the Mars Science Helicopter..In addition, the Mars fetch rover will be discontinued, and helicopters will be used to retrieve samples for the Mars Sample Return Mission. It will be called Fetch.

Mars Science Laboratory Curiosity

The Mars Science Laboratory (MSL) landed successfully on the surface on August 6, 2012. It had been launched on November 26, 2011. It's location on Mars is the Gale crater, and was a project of NASA's Jet Propulsion Laboratory. The project cost was around $2.5 billion. It is designed to operate for two Martian years (sols). The mission is primarily to determine if Mars could have supported life in the past, which is linked to the presence of liquid water. NASA has a Planetary Protection Officer assigned to Mars, to make sure we don't contaminate it with Earth life, and to protect any lifeforms that we may find there. So far, Earth is the only place we are aware of that has life. Is life on Earth unique in the Universe, or it is common? Mars may hold that answer for us.

Gale Crater is a location of interest. It's about 96 miles in diameter, and has a mountain, Aeolis Mons, in the center that is 18,000 feet high.

The Rover vehicle *Curiosity* weights just about 1 ton (2,000 lbs.) and is 10 feet long. It has autonomous navigation over the surface, and is expected to cover about 12 miles over the life of the mission. The platform uses six wheels. The Rover Compute Elements are based on the BAE Systems' RAD-750 CPU, rated at 400 mips. Each computer has 256 Mbytes of RAM, and 2 Gbytes of flash memory. The power source for the rover is a radioisotope thermal power system providing both electricity and heat. It is rated at 125 electrical watts, and 2,000 thermal watts, at the beginning of the mission. The operating system is WindRiver's VxWorks real-time operating system. The vehicle was assembled and tested at NASA's Goddard Space Flight Center, and shipped to lead center JPL. The landing location in Gale Crater was named Bradbury Landing, after the science fiction writer, Ray Bradbury. Mars figured heavily in his writings. Gale Crater is named after an Australian amateur astronomer, Walter Gale. There is some evidence that the crater was once filled with water.

The computers interface with an inertial measurement unit (IMU) to provide navigation updates. The computers also monitor and control the system temperature. All of the instrument control, camera systems, and driving operations are under control of the onboard computers.

Communication with Earth uses a direct X-band link, and a UHF link to a relay spacecraft in Mars orbit. At landing, the one-way communications time to Earth was 13 minutes, 46 seconds. This varies considerably, with the relative positions of Earth and Mars in their orbits around the Sun.

The science payload includes a series of cameras, including one on a robotic arm, a laser-induced laser spectroscopy instrument, an X-ray spectrometer, and x-ray diffraction/fluorescence instrument, a mass spectrometer, a gas chromotograph, and a laser spectrometer. In addition, the rover hosts a weather station, and radiation detectors. There is cooperation between in-space assets and ground rovers in sighting dust storms by the meteorological satellite in Mars orbit.
In 2013, NASA uploaded a software upgrade to Curiosity's operating System. Overall, it took a week to install.

Curiosity's exploration of the ancient lake bed, known as Gale Crater resulted in some new discoveries that NASA released on June 7, 2018. It found organic molecules, particularly methane, below the surface. Curiosity has a sampling drill (that, unfortunately, is limited to 5 cm.), a mass spectrometer, and a gas chromatograph. On Earth, most methane is from biological processes. It can be produced by in-organic processes, however. Scientists have also discovered a seasonal pattern in the amount of methane in the atmosphere, in the amount of a factor of three, that may point to sub-surface storage. Methane was first detected on Mars in 2003.

The source of the Mars methane has yet to be resolved. This will be a major goal of NASA's and ESA's next landers. A large underground reservoir of methane could be very useful for return-trip rocket fuel. You can convert methane to liquid methanol. Recall that Space-X ships use methane as rocket fuel. Space-X prototyped the process in 2011, producing one metric ton of oxygen and methane using 17 megawatts from solar power.

Maven

NASA's Maven mission to Mars is an orbiter. It is studying the Martian atmosphere It was launched in November of 2013, and reached Mars in September of 2014. Maven is an acronym for Mars Atmosphere and Volatile Evolution Mission. It was tasked with a one year mission in 2016, and was approved for an extended mission of another year. It has enough instruments working, and ehough fuel to operate until 2030.

MAVEN is equipped with a RAD-750 Central Processing Unit manufactured by BAE Systems in Manassas, Va. The processor can endure radiation doses that are a million times more extreme than what is considered fatal to humans. The RAD750 CPU itself can tolerate 200,000 to 1,000,000 rads. Also, RAD750 will not suffer more than one event requiring interventions from Earth over a 15-year period.

The RAD750 operates at temperatures of -55°C to 125°C with a power consumption of 10 Watts.

One discovery by Maven is that Mars' atmosphere deteriorates during solar storms. Maven can also serve as a communications relay for rovers on the surface. It is currently operational.

Insight

The Insight Mission launched in May of 2018, and landed on Mars in November of that year. It is a robotic lander with a seismometer and a heat transfer probe. It is based on the Phoenix design.

The mission landed near the Martian equator, which will provide maximum solar power during the 2-year mission.

The hope is, we can establish a refueling station on Mars or one of its moons, so we don't need to carry our return fuel with us. Since Mars has been proven to have water ice, this might be quite feasible. Using solar energy, you can break down water into liquid hydrogen and liquid oxygen – rocket fuel and oxydizer. NASA has published a map of water ice on Mars, and ESA's Mars Express spacecraft spotted the remails of an ancient river system and lake as well as Kraken crater, full of ice.

The explorer is still operating as of this writing.

Interplanetary Cubesat

Mars Cube One (MarCO) is the first interplanetary cubesat mission, headed by JPL. It involved sending two Cubesats to Mars, along with the Insight Rover. The 6U cubesats separated at Earth orbit and proceeded on their own. The mission was to be launched in March, 2016, when Mars was 1.07 AU distant, and arrived in September of 2016. The launch had been postponed due to a vacuum leak in the prime instrument. The mission was rescheduled for Spring, 2018. The mission officially ended on February 2, 2020, and was deemed successful.

A Cubesat is a small, affordable satellite that can be developed and launched by college, high schools, and even individuals. The specifications were developed by Academia in 1999. The basic structure is a 10 centimeter cube, (volume of 1 liter) weighing less than 1.33 kilograms. This allows multiples of these standardized packages to be launched as secondary payloads on other missions. A Cubesat dispenser has been developed, the Poly-PicoSat Orbital Deployer, P-POD, that holds multiple Cubesats and dispenses them on orbit. They can also be launched from the Space Station, via a custom airlock. ESA, the United States, Russia, and others provide launch services. The Cubesat origin lies with Prof. Twiggs of Stanford University and was proposed as a vehicle to support hands-on university-level space education and opportunities for low-cost space access. This was at a presentation at the University Space Systems Symposium in Hawaii in November of 1999.

The Cubesats serve as a real-time communications relay with Earth during the critical descent and landing phase of the rover. The lander talks to the Cubesat relays over an 8 kbps UHF link, and the Cubesats relay this to Earth over an 8 kbps X-band link to the DSN.

The Cubesats are stabilized with reaction wheels, and have propulsion systems to unload the wheels, and adjust their orbital position.

JPL is exploring a strange new rover, shaped like a white ball, 6 feet in diameter. It is currently being tested in Antarctica. It is wind driven. It doesn't get much choice where it goes, but it doesn't use fuel. It is called the tumbleweed rover. During testing, it did about 3.7 mph getting up to 13 at one point. It used a satellite link to send its location, the temperature, and other data. A similar terst was completed in Greenland. (As an aside, the author participated with a group of students at GSFC to develop and deploy Grover, the Greenland Rover. It was a tracked vehicle about the size of a small card, and carried an ince penetrating radar. It's over at the GSFC Visitor center).

The Russian/Soviet Mars Missions

Russian tried five Mars missions from 1960-1962. There were four launch failures, and the other lost communication before it reached the red planet. The same thing happened to another try in 1964. Three more launch failures came in 1969-1971. Finally, the Mars 2 and Mars 3 spacecraft reached and orbited in May of 1971, and operated for 362 orbits. These spacecraft deployed landers to the Martian surface. The small rovers were equipped with skis, and tethered to the lander with a umbilical of 15 meters. The Mars 2 became the first spacecraft to impact the Martian surface. It's location is unknown.

Zond-2 in 1964 was supposed to be a flyby, but communications were lost. Mission 2M 522 was launched in 1969, but had a launch failure. Kosmos 419 mad it to Earth orbit, but was stuck there. Mars mission 2M 521 in 1969 failed to achieve Earth orbit. Prop-M 's carrier vehicle failed before the rover could be deployed.

Mars-4, in 1973, was a partial success, returning photographs from orbit, but failed to perform its orbital insertion engine burn. Mars 5 worked for 9 days in orbit, and successfully returned some images, Mars 6 was a successful fly-by , but its lander lost contact after landing. Not giving up, Mars 7 was a successful mission. It's lander, unfortunately failed to enter the atmosphere.

Over twenty Russian/Soviet Mars spacecraft have been the victims of launch or payload failure. Mar-96 and its lander never left Earth orbit.

In the 1960's the Soviets wanted to send a crew to the Venus-Mars loop, without landing. The booster they were counting on never emerged.

Phobos-Grunt

In November of 2010, the Russian Space Agency launched an ambitious mission to set a probe down on the small Martian moon Phobos, collect samples, and return them to Earth.

There was a failure of the spacecraft propulsion system that stranded the mission in Earth orbit. It re-entered the Earth's atmosphere in January 2011, along with its rover.

Various causes were postulated for the failure, including interference by U.S. Radar, cosmic ray induced upsets, programming errors, and counterfeit chips.

The final report from Roscosmos cited software errors, failure of chips in the electronics, possibly due to radiation damage, and the use of non-flight qualified electronics, with inadequate ground testing.

Evidently, identical chips in two assemblies failed nearly simultaneously, so quickly that an error message was not generated. It was possible that the error was recoverable, as the spacecraft entered a safe mode with a proper sun orientation for maximum power. However, the design precluded the reset mode before the spacecraft left its parking orbit. This was major design oversight.

The identified chips that failed were 512k SRAM. The part numbers from the Russian report were checked by NASA's Jet Propulsion Lab, and were found to be among the most radiation susceptible chips they had ever seen. Bad choice. The chips could last in space a few days, and were barely acceptable for non-critical applications, The probably failure cause was single event latch-up, which is sometimes recoverable. In single event latch up, there is a single particle strike that latches up a transistor, preventing it from operating. Usually, if you turn it off and back on again, it will work. A lot of radiation damage to the underlying semiconductor lattice fixes itself after a while, a process called "annealing." Common practice is to use radiation-harden parts.

A second Phobos mission suffered a communications loss before the lander deployment and rover were deployed

ESA Missions

This section discusses the Mars missions of the European space agency. ESA has 22 member states, that participate and cooperate in missions and launches. It was founded in 1975. It operates a launch site in French Guiana on the east coast of South America. There is an ongoing study mission for human visitation, called Aurora.

Mars Express

This mission, in 2003, was a success, and remains so as of this writing. It is expected to remain operational through 2035. A lander, the U. K. Beagle2, was deployed, but its solar panels failed to deploy, ending its part of the mission. The orbiter uses solar panels for power. Due to adoption of standards for interplanetary communication, the spacecraft is inter-operable with NASA's Deep Space Net.

Rosetta

This mission was launched in 2004, and conducted a successful imaging flyby of Mars in 2007, en-route to its primary target, a comet. Its primary mission was the comet 67P/Churyumov–Gerasimenko.. It was part of the Horizon 2000 program.

The Indian Mars Mission

The Mangalyaan orbiter was launched in 2013, and entered Mars orbit in September of 2014. It is operational at this writing, and has an extended mission through 2020. India is the first nation to reach Mars successfully on its first attempt. The spacecraft spent about a month in Earth orbit, before beginning a trans-Mars injection maneuver. It is operated from the ISRO Control Center in Bangalore. It was lifted to orbit by the Indian (Polar Satellite Launch Vehicle) PSLV-XL launch vehicle. The spacecraft has three solar array panels, providing 840 watts of power at Mars.

The mission was planned for a 6-month duration, but is heading into its sixth year. The spacecraft weighs 1336 kg, and is a 1.5 meter cube. The mission has already exceeded expectations, came in under budget, and is returning significant new data about Mars. It has sent back two terabytes of data. Even China was impressed, calling the mission the pride of Asia.

A second mission is planned for the launch opportunity in 2024.

United Arab Emirates

The UAE Mars Al-Amal (Hope) mission was launched in 2020. It's mission is to study the Martian atmosphere, and, most importantly, bring a new player into the field. The Mission was announced in 2014, and was implemented by the UAE Space Agency, and the Mohamed bin Rashid Space Center.

The launch mass was 1,500 kg. It orbits the planet at altitudes between 22,000 and 44,000 km. The selected launch vehicle was Japanese, from Mitsubishi Heavy Industries, the H-IIA. The instrument suite includes an imager, an Infrared spectrometer, and an Ultraviolet spectrometer. It monitors daily and seasonal weather cycles, and dust storms.

It masses nearly 2,000 pounds and is the size of a small car.

Japan

The Nozomi (Planet-B) Mission in 1998 suffered a spacecraft failure. In 2024, Japan is planning an orbiter mission, MMX, Mars Moons Exploration, and an orbiter TEREX-1.

Melos is a Japanese rover concept, in study. Since the mission aims for access to a "special region", strict planetary protection sterilization protocols must be followed to prevent forward contamination of Earth microbes to Mars

Chinese Mars Mission

The Yinghuo-1 was deployed with the ill-fated Phobos-Grunt mission in 2011, and did not make it past Earth parking orbit. End of mission occurred in January of 2012, when it reentered the atmosphere, and burned. The Mission was a outgrowth of a Cooperative Agreement between the Russian and Chinese Space Agency's.

As part of their latest Mars Mission, Tianwen-1, China sent 6 units, an orbiter, two deployable cameras, a lander, and the Zhurong Rover. All together about 5 tons of equipment. It was launched in July, 2020, and the lander touched down in May of 2021.

The Rover drives down a ramp from the lander to the surface. It has ground penetrating radar, a magnetometer, a weather station, a spectroscope, a multispectrum camera, and a navigation camera. Early data from the south polar region seems to show the presence of water ice.

As part of the Eleventh Five-Year Plan, China is considering a crewed mission in the 2040 time frame.

Future Missions

The most exciting mission will be the crewed mission to Mars. In the meantime, there are more robotic exploration missions planned for the next launch opportunity. ESA is planning the ExoMars mission in partnership with Roscosmos.

This will use the Kazachok lander, and the Rosalind Franklin Rover. The rover is named after an English Chemist and crystallographer who mapped out the molecular structure of DNA.

India is planning their second Mars mission also.

Exo-MARSrosalind

Exo_Mars is an ESA astrobiology mission to search for evidence of present or past life on Mars. The first part was launched in 2016. This was a communications satellite in Mars orbit that can also do research on trace gases. The mission carried the Schiaparelli lander, which, unfortunately, crashed. The second part of the mission will put a lander and rover on the surface for exploration. It will look at water and trace gases, and also bio-signatures of possible Martian life. This lander, built by Roscosmos, will deploy the rover. The lander will do a controlled descent with parachutes and retro-rockets. The exobiology laboratory, named Pasteur, will look for signs of life.

The second part of the mission was due to launch in 2022. It carried the Rosalind Franklin rover. The mission was due to launch on a Russian rocket, but was suspended by the Ukraine crisis. NASA will now provide the rocket.

Mars-2020

This U. S. Program is a surface rover mission. It is designed to study astro-biology, surface geological processes and assess past habitability of the planet. It will search for bio-signatures in geological materials.

The Rover is a 6-wheeled derivative of the Curiosity design. It is a JPL mission. It was being discussed how the Rover could launch collected samples to Mars orbit, and how these could then be returned to Earth. The rover is accompanied by a drone helicopter, to scout out in front for potential areas of interest, and hazards.

Perseverance and Ingenuity

Perseverance, or Percy, is a automobile-sized rover, at the crater Jezero as part of NASA's Mars 2020 mission. It landed in February 2021. The site was named the Octavia E. Butler site, after a famous American science fiction writer. Perseverance's design follows that of the earlier Curiosity. There are seven science instruments, and nineteen cameras. It also has a mini-helicopter, Ingenuity. That craft weighs 1.8 kg. The rotors are 1.2 meters long, and it has four landing legs. It uses solar power. There are dual counter-rotating rotors for Mars' thin atmosphere. The blades have to rotate about 10 times faster than they would on Earth, since Mars' surface atmosphere is similar to Earth's at 27,000 meters. By February of 2022, it had made 19 flights.

The helicopter was situated under the Rover until landing, when it was dropped, and the Rover drove out of the way. It's initial take-off and landing area was named Wright Brothers Field.

It uses a Qualcomm Snapdragon 801 processor running linux. There are two flight microcontrollers. There were 150 engineers working on the project. It cost around $80 million. It uses navigation cameras. More advanced models are in development.

In 2024, the ESCAPADE Mars mission, consisting of two orbiters, is planned. The name is an acronym for Escape and Plasma Acceleration and Dynamics Explorer.

Planned Crewed Mars Missions

This section looks at the planned crewed missions to the Red Planet, from Space-X and NASA.

Space-X

Space-X, the Space Exploration Company, is a California Company doing launches and spacecraft manufacturing It was founded in 2002 by Elon Musk of Tesla fame. It has a constellation of commercial communications satellites in orbit, Starlink, and is involved in re-supplying and re-crewing the International Space Station. Space-X uses recoverable and reusable rockets. Musk donated $100,000 to the Mars Society. He has applied the modular and commercial off-the-shelf parts approaches to his vehicles, with great success in terms of cost and reliability. His rockets are recoverable and reusable. Elon Musk says he will have people on Mars by 2028. We shall see. I hope that's true.

NASA

NASA want to put astronauts on the surface of Mars by 2030. The United States has multiple robotic missions currently exploring Mars, with a sample-return planned for the future. The Orion Multi-Purpose Crew Vehicle (MPCV) is intended to serve as the crew delivery vehicle, with the Deep Space Habitat module providing additional living-space for the 16-month-long journey. The first crewed Mars Mission, orbiting Mars, and a return to Earth, is proposed for the 2030s. Technology development for US government missions to Mars is underway, but there is yet no well-funded approach to bring the conceptual project to completion with human landings on Mars by the mid-2030s, the stated objective. NASA is under presidential orders to land humans on

Mars by 2033 although later years like late 2030 seem more realistic. Unlike the Moon, we don't yet have a sample of the surface regolith yet.

Getting to Mars

Getting to Mars is not easy, and requires a lot of energy and time. First, we get to Earth orbit, and check that all the systems have survived launch. Knowing where Mars is currently, we can calculate the correct point to head to, when our mission reaches the correct distance. When the mission reaches Mars, we have to slow down and enter Martian orbit. Counter-intuitively, a minimum energy mission (but not a minimum time scenario) takes us looping past Venus to get to Mars in an efficient manner.

The distance from Earth to Mars varies on their relative positions in their orbits, and ranges from 33.9 to 250 million miles.

Every 26 months, the Earth and Mars align such that a minimum energy transfer can get a spacecraft from one to the other. The implications are that if you miss the launch window, you need to wait some 2 years, and that there is a minimum transit time for a Earth to Mars journey, landing, and return to Earth. This sets a mission duration that is important for future crewed missions. The travel time for this path is about nine months.

A Hohmann transfer orbit is an elliptical orbit between the circular orbits of two planets. It allows for the maximum payload to be sent with a given energy expenditure (or, amount of rocket fuel). You can get there quicker expending more energy. The alignment, or launch window, is critical. The details of the interplanetary transfer orbit were defined and published by German scientist Walter Hohmann in 1925.

The Hohmann transfer is a minimal energy approach. However it has a travel time of 9 months, a 500 day mission at Mars, then 9 months to return to Earth. A more energetic approach could be accomplished in 245 days.

Another trick is, instead of using fuel to slow down at Mars, we could employ aerobraking, even with Mar's thin atmosphere. Fuel needs to be expended for landing on the surface. Return fuel must be factored in, so we either have an accompanying tanker, or we get the in-situ methane facility going.

When we consider the synodic period, Earth-Mars trips are feasible every 26 months. There are also periods about every 15 years when the energy required for the trip is 50%. The next period is 2033, so look for some activity around that date.

The hope is, we can establish a refueling station on Mars or one of its moons, so we don't need to carry our return fuel with us. Since Mars has been proven to have water ice, this might be quite feasible. Using solar energy, you can break down water into liquid hydrogen and liquid oxygen – rocket fuel and oxidizer, in addition to the methane-oxygen approach.

Another use of a permanent colony that would bring in funds is space tourism, the Red Planet tour.

Getting around Mars

I put this out to the world-wide STEM community as a challenge. Design a viable railroad for Mars. We are going to need it eventually. It will take a while. Start now.

When humans arrive on Mars, and for a while afterwards, they will be completely dependent on periodic resupply missions from Earth. They will need to find their own supplies, chiefly water and methane, later, minerals and metals. Mars will be in the pre-Industrial Revolution phase, Just like when the early American settlers were dependent on resupply ships from Europe. Except the early American settlers found abundant food and indigenous people.

Railroading on Mars. This has not yet been done. It will be done eventually by someone.

Perhaps the closest railroad to what we might see on Mars is the National Railway of Mauritania. This is a traditional steel track on ties arrangement, and crosses more than 430 miles of the Sahara, from the iron mines at Zouerate (in the interior) to the port of Nouashibou on the Atlantic. The ambient temperature is around 120 degrees, F. Not in the shade, because there is no shade.

The locomotives are standard units from U. S. Manufacturer EMD, specified modified to operate in a sandy environment. The heat problem is addressed with additional insulation, and cooling systems. Dust and sand infiltration is the big problem. They have multiple sand plows, including one that can be moved by the engineer in the cab. Wind-blown drifts are an issue.

For desert operations, a pulse filtration system was added, as well as enhanced door and car body sealing. The cabin and body shell is pressurized. The gear cases have labyrinth sealing, so they can out-gas, but sand cannot not intrude on the moving parts. The units maintain their dynamic brakes. All fan blades receive a protective coating to counter sand erosion.

So, what is applicable to the Mars environment? Sand/dust plows, sealed gear cases, dynamic brakes, coated fan blades.

In the Martian environment, the cab would be accessed via an airlock, and the machinery would be mostly in a sealed environment, pressurized to just above Martian ambient. The big issue is cleaning off all the dust after entering the air lock, so it does not get in the cabin. This might be accomplished by over pressurizing the airlock, and then cycling it to the outside, before the crew member enters the cabin.

Dynamic brakes run the electric drive motors backwards, generating power as the train slows, and, on Earth, dissipating the heat via large resistor banks. A better approach is to follow the hybrid car scenario, where the energy is stored in batteries. This is being studied for terrestrial locomotives, but a very large battery would be required. The Mars locomotive would likely be a "hybrid" design with a large battery, to allow it to "limp home" if external power is lost.

Interplanetary Internet

Communications between planets in our solar system involves long distances, and significant delay. New protocols are needed to address the long delay times, and error sources.

A concept called the Interplanetary Internet uses a store-and-forward node in orbit around a planet that burst-transmits data back to Earth during available communications windows. At certain times, when the geometry is right, the Mars bound traffic might encounter significant interference. Mars surface craft communicate to Orbiters, which relay the transmissions to Earth. This allows for a lower wattage transmitter on the surface vehicle. Mars does not (yet) have the full infrastructure that is currently in place around the Earth – a network of navigation, weather, and communications satellites.

For satellites in near Earth orbit, protocols based on the cellular terrestrial network can be used, because the delays are small. In fact, the International Space Station is a node on the Internet. By the time you get to the moon, it takes about a second and a quarter for electromagnetic energy to traverse the distance. Delay tolerant protocols were developed for mobile terrestrial communication, but don't work in very long delay situations.

We have a good communications model and a lot of experience in Internet communications. One of the first implementations for space used a File Transfer Protocol (FPP) running over the CCSDS space communications protocol in 1996.

The formalized Interplanetary Internet evolved from a study at JPL, lead by Internet pioneer Vint Cerf, and Adrian Hook, from the CCSDS group. The concepts evolved to address very long delay and variable delay in communications links.

The Interplanetary Internet implements a Bundle Protocol to address large and variable delays. Normal IP traffic assumes a seamless, end-to-end, available data path, without worrying about the physical mechanism. The Bundle protocol addresses the cases of high probability of errors, and disconnections. This protocol was tested in communication with an Earth orbiting satellite in 2008.

The Artemis Project will be setting up a communications network on the moon to handle local communications, and communications with Earth. It will include wift.

Crewed Missions

Space-X wants to launch space tourists to Mars in the early 2020's. NASA is targeting getting astronauts near or on the surface of Mars by 2030. Mars Oasis was a 2001 project to land a experimental greenhouse with seeds that would be planted in hydrated martian soil. Things got going n 2012 when the Raptor engine for the Starship was in design.

The overall plan is to develop an extended human surface presence, to evolve into a self-sufficient colony. An initial group of a dozen or so is envisioned. To support this colony, we would need about 10 cargo flights per every human trip. Space-X Dragon spacecraft for a Mars

Mission will be called a Red Dragon.

Human Missions to Mars

Because of the relative positions of the Earth and Mars and the great distance between the two, a crewed flyby is probably not worth the effort. To place the first astronauts on Mars will involve building habitats and solar array farms, getting the vehicle re-fueled for the return journey, and just scoping out the place. Ultimately, with much work and much money, the long time goal is to establish a permanent colony, always occupied.

Talking about boots on Mars started in the early 1950's. It was postulated it might be feasible in the 1980's. Von Braun envisioned it in 1947. Both the U.S. And the Soviets worked on the concept. The NASA Space Exploration of 1989 got the momentum going. Design reference missions were formulated and revised in the 1990's. In 2001, ESA was working on their Aurora program, where one of the goals were crewed flight to look for traces of life beyond Earth. China's Space agency was also working on Mars missions.

Back in the 1950's, General Atomics was focused on a nuclear pulse propulsion spacecraft. The Nuclear Test Ban Treaty of 1963 killed this idea. Probably a good idea.

In 1962, General Dynamics and Lockheed Martin were doing design study's of Mars missions. Their concept involved eight Saturn-V flight to Earth orbit. Von Braun wanted to use Saturn-V's to launch a NERVA upper stage with a pair of spacecraft, each holding six crew. NASA went with the Shuttle instead.

Orion Capsule

The initial missions of Orion will be flights to the International Space Station and the Moon. The Orion capsule is reusable, and supports a crew of four. A first, un-crewed launch was successful in 2014 by a Delta-IV Heavy vehicle out of Launch Complex 37 at Cape Canaveral. The capsule was recovered at sea. The pressurized volume is 691 cubic feet, with a habitable volume of 316 cubic feet. The capsule has 50% more volume than Apollo. After it is proven on lunar missions, it will be the prime candidate for Mars travel.

Deep Space Gateway

The Deep Space Gateway (DSG) is a NASA Project for a crewed station in cis-lunar space. It is intended as a jumping-off point. The Orion crewed vehicle is scheduled to be used for this effort. The Gateway would be located in a halo orbit around the Moon. By that, we mean that the spacecraft would be visible to Earth for its entire orbital path. The DSG would form an in-orbit ecosystem to support missions to and on the lunar surface, and to Mars.

The DSG will result in a one-year crewed mission near the moon, to validate the concept of a flight to Mars. It is not so much the distance to Mars, as the relative orbital positions of the two planets in their solar orbits. The mechanics of the transfer orbit were worked out in 1925 by German scientist Walter Hohmann. In his 1928 book, *A Daring Trip to Mars,* Max Valier shows that one of the most efficient methods of reaching Mars from Earth involves a non-intuitive Venus fly-by. People have been thinking about this for a while. The shortest travel time occurs about every 26 years.

Using a standard Hohmann transfer orbit would involve a 9 month travel time, 500 days at Mars, and another 9 month return journey. Due to this time-frame, there is a significant radiation risk, both in space, and on the Martian surface. There is also the issue of lack of gravity for that period of time. Not only that, you could run out of potatoes.

Infrastructure

It is not enough just to go to Mars. The long term plans for a continuously crewed habitat, and medical facility's. This will allow, among other things, observation of other objects in our solar system from a different vantage point. In the long run, we might consider terra-forming Mars.

You won't be able to make a simple phone call back to Earth, without experiencing long delays.

Cost of transportation of goods from Earth is a limiting factor. We will want to prospect for things that we need that we won't need to bring from Earth. Radiation shielding is one. With the proper machinery, habitats can be buried beneath the surface. We will want to set up greenhouses to grow vegetables and fruits. These could also be a good return item and could bring large prices if shipped back to Earth. We know Mars has iron ore, and we know it has liquid water, What we really need is oxygen and return-trip rocket fuel.

Anything going to Mars will have to be sterilized inside and out. The same is true for the return missions.

Colonization of Mars

In the long run, we want a permanent colony. That's going to need a lot of money and a lot of work.

Mars Analog Bases on Earth

Several Mars analog bases have been constructed, and long term studies have been conducted for year. One of these is in the Canadian Arctic, and another in Utah, in the American west. The analog bases allow teams to operate as if they were on Mars. They were spacesuits while outside, and communicate over delayed radio with "mission control." Some missions have lasted a year or more.

A specific program addressing the Mars issues is the Flashline Mars Arctic Research Station (FMARS). There is currently one such facility in the Arctic, with a second planned. The existing station is on Devon Island in the Arctic sea. It is located on a ridge, overlooking a large impact crater, about a thousand miles from the North Pole. The facility was built in 2000, and is operated by the Mars Society, a non-profit. It is used to define and refine field procedures, test habitat design, study crew performance, and selection criteria. It began operations in 2001. Generally, there is a core crew of ten, with visiting researchers and assistants. A hazard probably not found on Mars is the occasional polar bear, looking for a quick meal. One outside crew member is always armed. Communication to and from the station to external sources is delayed 20 minutes, to simulate the one-way radio/light travel time to Mars. The crew keeps to the somewhat longer Martian Sol day

The Biosphere -2 project, located in the Arizona desert, supports 8 humans for a year in a closed ecosystem.

The Chinese have a Mars Analog Base on the high planes of Tibet.

Mars Base Camp

When we send the first human mission to Mars, it would be nice if they could refuel there for the return trip. There is water ice below the surface. It is also possible that Mars' tiny moons contain water ice. Why water ice? You use solar power to break it down into hydrogen and oxygen – the perfect rocket fuel. A big advantage is that you don't need to carry the return fuel on the outgoing flight.

Human missions to Mars have gotten consideration for more than 150 years. Von Braun made a very detailed study of a Mars mission in 1952. Willy Ley had published a variation in 1949. By 1956, their modified mission design would required over 400 launches, assembly in space, but would provide a winged lander for Mars. The thought is to land there, and, now that we know the surface features, begin to terra-form the planet for our needs. Mars missions are in active planning by the United States, Russia, Europe, China, and several commercial entities. Since Mars and the Earth have different orbital periods ("years") around the Sun, there are optimal times to make the journey. After working through all the math, the time between optimal Earth-Mars trips is 26 months. This provides the optimal energy expenditure. Using a Hohmann transfer, there would be a 9-month travel time, Earth to Mars, a 500 day stay at Mars to allow alignment of the orbits again, and another 9 month journey back. Can't we do better? Yes, but at the cost of fuel. There is a maneuver that would provide a Venus and a Mars flyby in one mission, with no landings. All of these long duration missions offer various hazards, and will be operating in "unknown territory" in terms of a group of humans isolated in a small space for a long time.

A NASA study by three major aerospace companies in 1962, showed a Mars mission requiring 8 Saturn-V "moon rockets," with assembly in Earth orbit. Von Braun's Mars Mission was passed over in favor of the Space Shuttle Project. Mars has been studied by fly-by, orbiting, and lander missions since 1981. NASA's Mars Design Reference Mission of the 1990's assumed that fuel could be synthesized from Martial atmospheric or surface components. The design studies continue to this day.

The Mars Base Camp is a crewed Mars orbiter, proposed by Lockheed Martin, possibly ready for the 2028 favorable launch opportunity. The hardware would use the Orion MPCV. Although the humans would remain in Mars orbit, they would perform tele-operation experiments on the surface. The mission would be launched from lunar orbit. The first mission is named Mars Base Camp-1. A crew of 6 would spend a year in Martian orbit. Interplanetary missions generally provide very limited options for abort/early return scenarios.

Lockheed defined a road map of the technologies required to achieve the Mission. The parts include: the Multi Purpose Crew Vehicle (MPCV), which is the orbital command and control center, implementing navigation, communications, life support and habitat systems. The Solar Array subsystem provides power for the system, including the electric propulsion engines. This is being developed by the NASA-Glenn Center. Radiators control dumping of excess heat. The propulsion stage uses cyrogenic fuels. Side visits to the moons Phobos and Deimos are planned. The main part of the assembly will consist of an Orion capsule with an associated service module, and an excursion module. There will be a laboratory and workshops. The excursion modules will provide access to the Martian (and Mars lunar) surfaces.

The Mars hardware and operations would be checked out at the Deep Space Gateway. The eventual goal is to establish a self-sustaining colony, and perhaps terra-form the Red Planet to be friendlier to Earth life, plant and animal.

The MPCV would provide housing, life support, transportation, and command & control. Large solar arrays would be used, as the vehicle is designed for electric propulsion. In addition, an onboard 3-D printer is seen as an alternative to manifesting spares.

A follow-on concept is a winged Mars surface lander, using liquid hydrogen and liquid oxygen engines. The fuel and oxidizer would be derived from water, electrolized using solar energy. Initially, the water would be shipped from Earth or the Moon, with a goal of processing water on the Martian surface, or its moons. NASA sees an opportunity for commercial firms to supply the water. A Water Delivery Vehicle (WDV) with a capacity of over 50 metric tons would be required. This would dock with a 375 kilowatt electrolysis plant in orbit. The Earth-rated 375 kw plant degrades to 160kW (42.6%) at Mars' greater distance from the Sun.

Modules would be delivered and pre-positioned in Mars orbit, including a lander (The Mars Ascent/Descent Vehicle) and a cyrogenic fuel depot. The MADV touches down and launches vertically. It will house 4 crew for a 10-day duration on the surface. The MADV's engines will use liquid hydrogen and oxygen for an efficient Isp of 405 seconds. Six RL-10 class engines are postulated.

The lander will have crew accommodations for four, on three decks. There is to be a flight deck, crew quarters on the mid deck, and an aft deck with galley, lab facilities, and the airlock.

Solar electric propulsion using xenon gas will be feasible with large solar arrays. The Deep Space Gateway will use this approach for station-keeping.

One timeline shows a cis-lunar outpost starting in 2020's, with lunar surface science by 2024. There could be a predeployment of assets to Mars by 2026. The year 2028 is targeted for the Mars mission. Besides the government mission, certain private spacecraft companies are defining their own Mars missions.

Commercial ventures such as Elon Musk's SpaceX are also interested in a Mars Colony. Musk, the founder of PayPal, put his own money into the company. It has developed its own launch vehicle, the Falcon, and has a 12-trip resupply contract for the ISS. SpaceX's approach to the uncrewed resupply missions is unique. The capsule as well as the rocketd are recovered and reused, saving a lot of money.

There are two 3D printers on the ISS, made by a California Company, Made-in-Space. NASA is exploring having the capability of printed custom parts in situ, as opposed to carrying spares, as a key for the Mars mission. Made-in-Space sees the orbital facility as an ideal location to manufacture optical fiber. Currently, items are printed for use onboard the station, as opposed to sending them up on a resupply flight, and other items are printed and returned to Earth for testing. A practical device that was printed was a custom buckle for exercise equipment, designed by Astronaut Yvonne Cagle.

Mars Crewed Fly-by

In 2013, the Inspiration Mars Foundation, a non-profit founded by Entrepreneur Dennis Tito, proposed a manned mission for a MARS fly-by for 2018. The details have subsequently been removed from the Website. This was estimated to require up to $2 billion ($10^9$) dollars. Space-X had been contacted as a possible vehicle provider. A "Plan-B" was defined using the known Venus-Mars flyby, in 2021 (when the planets were aligned properly).

There is also a one-way trip option proposed by several space enthusiasts. This option assumes you get it right the first time, and there are still a lot of unknowns. On the other hand, establishing a colony may be easier than the return journey. The current thinking is to launch the habitats and required equipment before the crew. It is still a leap of faith.

Afterword

Mars. We;re going there. It will be hard, and there is a lot to learn. Once we are established on Mars, we can explore and exploit the asteroids. We will have a jumping off point for exploration of the icy moons of Jupiter and Saturn. It's going to be interesting.

Bibliogaphy

Adler, Mark; Owen, W.; Riedel, J. "Concepts and Approaches for Mars Exploration, 2012
avail:
https://www.lpi.usra.edu/meetings/marsconcepts2012/pdf/4337.pdf

Aldrin, Buzz; David, Leonard *Mission to Mars: My Vision for Space Exploration,* 2015, ISBN-1426214685.

Aldrin, Buzz; Dyson, Marianne *Welcome to Mars: Making a Home on the Red Planet*, 2015, ISBN-1426322062.

Atreya, Sushil K.; Mahaffy, Paul R.; Wong, Ah-San (2007). "Methane and related trace species on Mars: origin, loss, implications for life, and habitability".Planetary and Space Science.55(3):, avail:

Baker, David *NASA Mars Rovers Manual: 1997-2013 (Sojourner, Spirit, Opportunity and Curiosity) (Owners' Workshop Manual), 2013,* ISBN-0857333704.

Bell, Jim *Postcards from Mars: The First Photographer on the Red Planet,* Dutton Adult, 2016, ISBN-0525949852.

Betancourt, Mark "CubeSats to the Moon (Mars and Saturn, too)", Air & Space Magazine, Sept 2014.

Bradbury, Ray *The Martian Chronicles,* ISBN-006207993X.

Burns, Jack O. Terry Fong NASA Ames Research Center David A. Kring William D. Pratt and Timothy Cichan, "SCIENCE AND EXPLORATION AT THE MOON AND MARS ENABLED BY SURFACE" TELEROBOTICS" INTERNATIONAL ACADEMY OF ASTRONAUTICS 10th IAA SYMPOSIUM ON THE FUTURE OF SPACE EXPLORATION: TOWARDS THE MOON VILLAGE AND BEYOND, Torino, Italy,

June 27-29, 2017.

Burt, Dennis *Elon Musk will Take Us to Mars: How and Why the Billionaire Entrepreneur and his SpaceX Start-Up are Making Interplanetary Travel a Reality,* 2013, ASIN-B00G0T6D6U.

Carney, Elizabeth *Mars, The Red Planet, National Geographic Kids,* ISBN-978-1-4263-2754-4.

Cichana, Timothy; O'Dell, Sean; Richey, Danielle; Bailey, Stephen A.; Burche, Adam "MARS BASE CAMP UPDATES AND NEW CONCEPTS," 2017, IAC-17, 68th International Astronautical Congress (IAC).

David, Leonard *Mars: Our Future on the Red Planet,* 2016, ISBN-426217587.

David, Leonard and Aldrin, Buzz *Mission to Mars: My Vision for Space Exploration,* 2013, ISBN-978-1426210174.

Editors of National Geographic; Daniels, Patricia *National Geographic Mars: Secrets of the Red Planet,* 2018, ISBN-1547842466.

Fischer, Maria "Mothership and her Hedgehogs: New Concept for Exploring Phobos" Space Safety Magazine, 2013, avail: http://www.spacesafetymagazine.com/space-exploration/deep-space/mothership-hedgehogs-concept-exploring-phobos/

Harland, David M. *Mars Owners' Workshop Manual: From 4.5 billion years ago to the present (Haynes Manuals)*, ISBN-1785211382.

Hohmann, Walter *THE ATTAINABILITY OF HEAVENLY BODIES,* Reprinted as NASA TT-F-44, 2017.
avail: http://large.stanford.edu/courses/2014/ph240/nagaraj1/docs/hohmann.pdf

Horneck, Gerda "General human health issues for Moon and Mars missions: Results from the HUMEX study" 2006, Advances in Space Research. 37 (1): 100–108. Bibcode:2006AdSpR..37..100H. doi:10.1016/j.asr.2005.06.077.

Kaufman, Mark *Mars Up Close: Inside the Curiosity Mission,* 2014 National Geographic, ISBN-142621278X.

Lakdawalla, Emily *The Design and Engineering of Curiosity: How the Mars Rover Performs Its Job,* 2018, Springer, ISBN- 3319681443.

Lane, Melissa D. "Hemitite on Mars: What does it tell us?," PSI Newsletter, Winter 2002, Vol 3, No. 4

Ley, Willy and Von Braun, Werner, *The Exploration of Mars*, 1956, Sidgwick & Jackson; First Edition, ASIN-B0000CJKQN.

Lowell, Perciva, *Mars and its Canals,* ASIN-B00U3ZRER4.

Lowell, Percival, *Mars: Is There Life On Mars?*, ISBN-1605065528.

Lowell, Percival, *Mars,* ASIN-B004ISKE6K.
Lowell, Percival, *The Evolution of Worlds,* ASIN-B00MSC2PJW.

Lowell, Percival, Mars as the abode of life, ASIN-B00JG57GA2.

Mackenzie, Bruce "To Mars - a Permanent Settlement on the First Mission," presented at the 1998 International Space Development Conference, May 21–25, Milwaukee WI.

McGinty, Michael *Canals on Mars: Lowell Revisited*, 2016, ASIN-B01BCD66BK.

NASA, *NASA's Constellation Program: Lessons Learned (Volume I and II) - Moon and Mars Exploration Program - Ares Rockets and Orion Spacecraft,* avail: http://www.thebookishblog.com/nasa-s-constellation-program-lessons-learned-volume-i-and-ii.pdf

NASA, *NASA Report on Mars Exploration: Frontier In-Situ Resource Utilization for Enabling Sustained Human Presence on Mars - ISRU, Surface Habitats, Entry*

Descent and Landing, Fuels, Food, Robotics, 2016, ASIN-B01JDORX58.

NASA, *NASA Space Technology Report: Lunar and Planetary Bases, Habitats, and Colonies, Special Bibliography Including Mars Settlements, Materials, Life Support, Logistics, Robotic Systems,* ASIN-B00CLX44E2.

NASA, *NASA Space Technology Report: Deep Space Habitat Concept of Operations for Transit Mission Phases - Mars, Phobos / Deimos, Near Earth Asteroid, Habitats, Crew Systems,* 2013, ASIN-B00EG4N3E6.

NASA, Human Exploration of Mars: Design Reference Architecture 5.0, 2014, ISBN-978-1495919961.

National Geographic, "Mars: Inside the High-Risk, High Stakes Race to the Red Planet, November 2016, https://www.nationalgeographic.com/magazine/2016/11/#

Paris, Antonio *Mars: Your Personal 3D Journey to the Red Planet,* 2018, The Center for Planetary Science, ISBN-0692073671.

Perminov, V.G. (July 1999). The Difficult Road to Mars - A Brief History of Mars Exploration in the Soviet Union. NASA Headquarters History Division. ISBN 0-16-058859-6.

Petranek, Stephen *How We'll Live on Mars*, 2015, Simon & Schuster, ISBN-1476784760.

Portree, David S. F. *Humans to Mars: Fifty Years of Mission Planning, 1950–2000*, NASA Monographs in Aerospace History Series, Number 21, February 2001, NASA SP-2001-4521. Avail: ASIN-B014RGH7GM.

Price, Hoppy; Hawkins, Alisa; Radcliff, Torrey *Austere Human Missions to Mars,* 2009, AIAA Space 2009 Conference.

Pyle, Rod *Destination Mars*. Prometheus Books. 2012, ISBN 978-1-61614-589-7.

Rapp. Donald *Human Missions to Mars: Enabling Technologies for Exploring the Red Planet,* 2015, Springer Praxis, ISBN-3319222481.

Sparrow, Giles *MARS* (Illustrated), 2015, ISBN-162365856X.

Taylor, Fredric *The Scientific Exploration of Mars*. Cambridge: Cambridge University Press, 2010, ISBN-978-0-521-82956-4.

Tolker-Nielsen, Toni "Exomars 2016 – Schiaparelli Anomaly Inquiry," 2017, ESA, DG-I/2017/546/TTN.

United States Congress and United States House of Representatives, *Next step to Mars: deep space habitat* : hearing before the Subcommittee on Space, 2017.

Valier, Max; Miller, Ron (ed), *A Daring Trip to*

Mars,1928, reprint, 2013, ASIN-B00CSWANK0.

Von Braun, Wernher *The Mars Project*, 1962, U. Illinois Press, ISBN-0252062272.

Von Braun, Wernher *Project MARS: A Technical Tale, 2006,* ISBN-0973820330.

Von Braun, Wernher; Ley, Willy *The Exploration of Mars,* 1956, ASIN : B0000CJKQN.

Wallace, Alfred Russel *Is Mars habitable? A critical examination of Professor Percival Lowell's book "Mars and its canals," with an alternative explanation,* ASIN-B0084C22WK.

Weintraub, David *Life on Mars: What to Know Before We Go*, 2018, ASIN-B079P8S585.

Wells, H. G. *The War of the Worlds,* ASIN-B07DFSCGZH.

Zubrin, Robert *Mars on Earth: The Adventures of Space Pioneers in the High Arctic,* 2003, ISBN-158542255X .

Zubrin, Robert *Entering Space: Creating a Spacefaring Civilization*, 2000, ISBN-10-1585420360.

Zubrin, Robert *Mars Direct: Space Exploration, the Red Planet, and the Human Future: A Special from Tarcher/Penguin,* 2013, ASIN-B00AMOO98I.

Resources

- NASAspaceflight.com
- http://www.nasa.gov/mars
- https://www.nasa.gov/feature/deep-space-gateway-to-open-opportunities-for-distant-destinations
- http://www.lockheedmartin.com/us/ssc/mars-orion.html
- Mars Base Camp, http://lockheedmartin.com/us/ssc/mars-orion.html
- "Mars Base Camp Updates and New Concepts" available for download at the address above.
- NASA's Exploration Systems Architecture Study -- Final Report, avail:

https://www.nasa.gov/exploration/news/ESAS_report.html

- Mars Foundation - https://www.space.com/19982-private-mars-mission-gallery-inspiration.html
- Human Exploration of Mars, Reference Mission avail: https://web.archive.org/web/20070626154441
- http://exploration.jsc.nasa.gov/marsref/contents.html
- Maps are available at Google Mars.
- A. C. Clarke - https://futurism.media/space-exploration-developments-by-2050-a-fictional-vision
-
- https://www.sciencedirect.com/science/article/pii/S

0032063306001814?via%3Dihub
- https://www.nasa.gov/feature/jpl/nasas-treasure-map-for-water-ice-on-mars
- wikipedia, various.

References

Pathfinder

http://www.nasa.gov/mission_pages/mars-pathfinder/
http://research.microsoft.com/en-us/um/people/mbj/Mars_Pathfinder/

MSL -Curiosity

www.nasa.gov/msl/
http://en.wikipedia.org/wiki/Mars_Science_Laboratory
www.space.com/16385-curiosity-rover-mars-science-laboratory.html
http://www.windriver.com/announces/curiosity/Wind-River_NASA_0812.pdf

Mars Climate Orbiter

http://mars.jpl.nasa.gov/msp98/orbiter/

References Phobos-Grunt
Klotz, Irene "Programming Error Doomed Russian Mars Probe," Discovery News, Feb. 7, 2012, news.discovery.com

de Carbonnel, Alissa "Russia races to salvage stranded

Mars probe, " Reuters, 2011. www.reuters.com

Amos, Jonathan "Phobos-Grunt mars Probe loses its way just after launch," 9 Nov. 2011, BBC News, www.bbc.co.uk

Oberg, James "Did Bad memory chips Down Russia's Mars Probe?," Feb 2012, IEEE Spectrum, IEEE.org.

Friedman, Louis D. "Phobos-Grunt Failure Report Released," 2/6/2012, www.planetary.org/blogs/guest-blogs/lou-friedman

Phobos fail: What really happened to Russia's Mars Probe, Jan 19, 2012, RT.com.

Specific References

Viking

https://www.nasa.gov/mission_pages/viking

http://www.maniacworld.com/Viking-Mission-to-Mars.htm (video)

The Viking Mission to Mars, NASA SP-334
https://ntrs.nasa.gov/archive/nasa/casi.ntrs.nasa.gov/19740026174.pdf

Mars Observer

https://nssdc.gsfc.nasa.gov/nmc/masterCatalog.do?sc=1992-063A

Mars Global surveyor

mars.jpl.nasa.gov/mgs/

Mars Odyssey

https://mars.jpl.nasa.gov/odyssey/index.cfm

Mars Exploration rover

https://mars.jpl.nasa.gov/mer/home/index.html

MRO

https://photojournal.jpl.nasa.gov/spacecraft/MRO

https://mars.jpl.nasa.gov/mro/

https://www.nasa.gov/mission_pages/MRO/main/index.html

Phoenix

https://www.webcitation.org/5W4NeGhno?url=http://phoenix.lpl.arizona.edu/

Maven

https://mars.jpl.nasa.gov/programmissions/missions/present/maven/
https://www.nasa.gov/mission_pages/maven/main/index.html

Insight

https://www.nasa.gov/mission_pages/insight/main/index.
https://mars.nasa.gov/insight/

Emirates Mars Mission

http://www.emiratesmarsmission.ae/mission-journey

ExoMARS

http://exploration.esa.int

http://exomars.cosmos.ru

Mars 2020

http://mars.jpl.nasa.gov/mars2020/

NASA'a map of water on Mars

https://www.space.com/mars-water-ice-map.html

wikipedia, various

Glossary of terms and definitions

Areography – equivalent of geography on Earth.
Apogee – furthest point in the orbit from the Earth.
Apoareion – farthest point in orbit, from Mars.
ARED – a resistive exercise equipment on the ISS
ASIN – Amazon Standard Inventory Number
Astrionics – electronics for space flight.
AU – astronomical unit.
Baud – measure of data rate; bits per second
BEM – bug-eyed monster
BEO – beyond Earth orbit.
Byte – data structure of 8 bits.
CBM – common berthing mechanism.
CCM – Contingency Consumables Module
CCS – command computer system
CCSDS – Consultive Committee on Space Data Systems, a standards organization.
CEV (Orion) Crew Exploration Vehicle
CM – crew module
CME – Coronal Mass Ejection, blast of energetic particles from   the Sun.
CMP – co-manifested payload.
CNSA – China National Space Administration.
Conops – concept of operations.
COTS – Commercial off-the-shelf; Commercial Orbital Transportation System.
CPS – Cyrogenic Propulsion Stage.
CPU – central processing unit
CRTBP – Circular Restricted three-body Problem.
CSA – Canadian Space Agency, Agence Spatiale

Canadienne
CSF – Cislunar Support Flight.
C&W – caution and warning.
Cygnus – Orbital-ATK automated cargo vehicle for ISS.
Cyrogenic – relating to very low temperatures.
DAM – damage avoidance maneuver.
DAV – Descent Ascent Vehicle
DCM – docking cargo module.
Delta-V – change in velocity.
DIPS – Dynamic Isotope Power System
DoD – (U.S.) Department of Defense.
DRA – Design Reference Architecture
DRG – Distant Retrograde Orbit.
DRM – design reference mission.
DSG – Deep Space Gateway
DSH – deep space habitat.
DSN – (NASA) Deep Space Network.
DST – Deep Space Transport, Mars Transit Vehicle
DTM – dynamic test model, for structural tests. Digital terrain models.
ECLSS – Environmental Control & Life Support system.
ECM - electronic core module.
Ecopoiesis – initiating life in a new place.
EDL – Entry, Descent, Landing
EDS – Earth Departure Stage.
Emirs - Emirates Mars Infrared Spectrometer.
EMUS - Emirates Mars Ul;traviolet Spectrometer.
EM-x Exploration Mission number-x.
EMM – Emirates Mars Mission.
Ephemeris – position information data set for orbiting bodies, 6 parameters plus time.

Epoch – a reference point in time for orbital elements.
EPS – electrical power system
ESA – European Space Agency
EUS – Exploration Upper Stage.
EVA – extra-vehicular activity.
EXI - Emirates exploration imager.
Flash – a type of non-volatile memory
FMARS – Flashline Mars Arctic Research Station
FPGA – Field Programmable Gate Array – an integrated circuit
FTL – faster than light
FTP – file transfer protocol.
G – one Earth normal gravity; as a prefix, 10^{12}
GCSC – Guidance, control, sequencing.
GHZ - giga-Hertz
GNC – Guidance, Navigation, and Control.
Gravity well – a conceptual model of the gravity field near a mass.
GSFC – NASA Goddard Space Flight Center, Greenbelt, MD.
GSLV - (India) Geosynchronous Satellite Launch Vehicle
GYRO – sensor for orientation.
Halo Orbit – three dimension orbit near the L1, L2, or L3 Lagrange points.
HAMO - High-Altitude Mapping Orbit
HEEO – highly eccentric Earth orbit.
Hemitite – an iron ore Fe_2O_3. It's red.
HEOMD – Human Exploration and Operations Mission Directorate.
HiRISE - High-Resolution Imaging Science Experiment.
HITL – Human in the loop.

HOPE – Human Outer Planet Exploration (NASA)
HSIR – human systems integration requirements
IDSS – International Docking System Standard.
IGA - (ISS) Inte-Governmental Agreement
IMU – inertial measurement unit.
IP – internet protocol
ISP – specific impulse. Measure of efficiency of rocket engine. Units of seconds.
ISRO – Indian Space Research Organization.
ISRU – in situ resource utilization.
ISS – International Space Station.
JAXA – Japan Aerospace Exploration Agency.
KW – kilowatt.
IDSN – Indian deep space network.
ISRU – in site resource utilization.
ISS – International Space Station
JAXA – Japanese space agency
JPL – Jet Propulsion Laboratory, Pasadena, CA.
JSC – Johnson Space Center, Houston, Texas.
K, kilo - 10^3
KSC – NASA Kennedy Space Center, launch site, Florida.
L2 – second of 5 Lagrange points, a null in the gravity field in the restricted 3-body problem.
LAS – launch abort system.
LASP - Laboratory for Atmospheric and Space Physics (Boulder, CO.)
Lbf – pounds, force.
LCH_4 – liquid methane.
LCT – Lunar Cargo Transportation.
LEO – Low Earth Orbit

LH$_2$ – liquid hydrogen.
Libration point – null in the gravity field of the three body problem.
Lidar - laser imaging, detection, and ranging.
LMO – low Mars orbit.
LOS – Russian Lunar Orbital Station; loss-of-signal.
LOX – liquid oxygen, boils at -297 F.
LSAM – lunar surface access module
LSPPO – Lunar Surface systems Project Office (NASA-JSC).
LST – landing by soft touchdown.
MADV – Mars Ascent/Descent Vehicle.
MARPOST – Mars Piloted Orbital Station.
MAV – Mars ascent Vehicle.
MAWG – Mars Architecture Working Group
MBC – Mars Base Camp.
MBRSC - Mohammed bin Rashid Space Centre (Emirates).
MELOS - (Japan) Mars Explorations with Landers and Orbiters" or Mars Exploration with Lander-Orbiter Synergy.
MER – Mars Exploration rover
MET – mission elapsed time.
MHS – Mars Helicopter Scout.
MHz – mega (10^6) hertz
MIPS – millions of instructions per second.
MMH – Monomethylhydrazine, CH3(NH)NH2
MMSEV – MultiMission Space Exploration Vehicle.
MOI – Mars orbit insertion
MOU – memorandum of understanding.
MPCV - Multi-Purpose Crew Vehicle.

MPH – miles per hour.
MPK – Martian Piloted Complex
MPLM – Multi-purpose Logistics Module.
MRO – Mars Reconnaissance Orbiter
m/s – meters per second.
MSL – Mars Science Laboratory
MSR – Mars sample return
MT – metric ton, 1000 kg.
N – Newton, metric unit of force.
NAC – NASA Advisory Council.
Nadir – the point directly below.
NASA – (U.S.) National Aeronautics and Space Administration
NEO – near Earth object.
NextSTEP-2 – (NASA) Next Space Technologies of Exploration Partnerships.
NHV – net habitable volume.
NRHO – Near rectilinear halo orbit (around the L1 or L2 Earth-Moon libration point).
NTIS – National Technical Information Service (www.ntis.gov).
Nto – nitrogen tetroxide.
NTR – Nuclear thermal rocket
NTRS – NASA Technical Reports Server, ntrs.nasa.gov
ORU – Orbital Replacement Unit.
OPSEK – (Russian) Orbital Piloted Assembly and Experiment Complex.
OWLT – one-way light time.
RAD – unit of radiation
Perigee – closest point in the orbit from the Earth.
Peri-areion – closest point in orbit to Mars.

PMA – Pressurized mating adapter.
PMCU – Power Management Control Unit.
PPB – power and propulsion bus.
PPO – Planetary protection officer.
PSI – Planetary Science Institute, Arizona, US
PTCS – Passive thermal control system
PVCU – Photo Voltaic Control Unit.
R&D – research & development.
RAM – random access memory
RCS – reaction control system.
Regolith – layer of loose material, covering rock; dirt.
RGA – rate gyro assembly
ROSCOSMOS – Russian Space Agency.
RPOD – Rendezvous, Proximity Operations, Docking.
RTE - return to Earth.
RTG – Radioisotope Thermoelectric Generator.
SCRAM – emergency reactor shutdown
SEP – solar electric propulsion.
SEU – single event upset, transient error in a digital circuit, usually due to radiation.
SHFE – space human factors engineering.
SI – System International – the metric system.
Sidereal period – time for an object to make a full orbit.
Sol, local solar day – on Mars, 24h, 37 min.
SLS – (NASA) Space Launch System.
SPACE Act - Spurring Private Aerospace Competitiveness and Entrepreneurship.
Sol – a local day.
SPACE-X – private space company.
SPLM – Surface Power and Logistics Module.
SRAM - static random access memory.

SRB – Supersonic Retro Propulsion
SurfHab – Surface Habitation
Synodic period - time for an object in orbit to occupy the same point, in relation to 2 other objects.
T – metric ton, 1,000 kg
TCS – thermal control system.
TEI – Trans Earth Injection
Tera – 10^{12}
TGO - Trace Gas Orbiter.
TLI – Trans-lunar injection.
TM – Technical Manual.
TMI – Trans Mars Injection
TOP – Trajectory Optimization Program
TransHab – Mars Transit Habitat.
TPS – thermal protection system.
Trillion - 10^{12}
TRL – technology readiness level.
UAE – United Arab Emirates
UAESA - United Arab Emirates Space Agency.
UDM – universal docking module.
UHF – ultra high frequency, 300 MHz to 3 Ghz
Ullage – residual fuel or oxidizer in a tank after engine burn is complete.
USAF – United States Air Force.
USGS - United States Geologic Survey
V&V – verification and validation.
WDV – water delivery vehicle.
VSTB – vehicle system test bed
X-band – 8 – 12 GHz.
XBASE - Expandable Bigelow Advanced Station Enhancement.

Zenith – the point directly above.
Zombie-sat – a non functional satellite in orbit, contributing to the orbital debris problem.

If you enjoyed this book, you might also be interested in some of these.

Stakem, Patrick H. *16-bit Microprocessors, History and Architecture*, 2013 PRRB Publishing, ISBN-1520210922.

Stakem, Patrick H. *4- and 8-bit Microprocessors, Architecture and History*, 2013, PRRB Publishing, ISBN-152021572X,

Stakem, Patrick H. *Apollo's Computers,* 2014, PRRB Publishing, ISBN-1520215800.

Stakem, Patrick H. *The Architecture and Applications of the ARM Microprocessors,* 2013, PRRB Publishing, ISBN-1520215843.

Stakem, Patrick H. *Earth Rovers: for Exploration and Environmental Monitoring,* 2014, PRRB Publishing, ISBN-152021586X.

Stakem, Patrick H. *Embedded Computer Systems, Volume 1, Introduction and Architecture*, 2013, PRRB Publishing, ISBN-1520215959.

Stakem, Patrick H. *The History of Spacecraft Computers from the V-2 to the Space Station*, 2013, PRRB Publishing, ISBN-1520216181.

Stakem, Patrick H. *Floating Point Computation*, 2013, PRRB Publishing, ISBN-152021619X.

Stakem, Patrick H. *Architecture of Massively Parallel Microprocessor Systems*, 2011, PRRB Publishing, ISBN-

1520250061.

Stakem, Patrick H. *Multicore Computer Architecture,* 2014, PRRB Publishing, ISBN-1520241372.

Stakem, Patrick H. *Personal Robots*, 2014, PRRB Publishing, ISBN-1520216254.

Stakem, Patrick H. *RISC Microprocessors, History and Overview,* 2013, PRRB Publishing, ISBN-1520216289.

Stakem, Patrick H. *Robots and Telerobots in Space Application*s, 2011, PRRB Publishing, ISBN-1520210361.

Stakem, Patrick H. *The Saturn Rocket and the Pegasus Missions, 1965,* 2013, PRRB Publishing, ISBN-1520209916.

Stakem, Patrick H. *Visiting the NASA Centers, and Locations of Historic Rockets & Spacecraft,* 2017, PRRB Publishing, ISBN-1549651205.

Stakem, Patrick H. *Microprocessors in Space*, 2011, PRRB Publishing, ISBN-1520216343.

Stakem, Patrick H. Computer *Virtualization and the Cloud,* 2013, PRRB Publishing, ISBN-152021636X.

Stakem, Patrick H. *What's the Worst That Could Happen? Bad Assumptions, Ignorance, Failures and Screw-ups in Engineering Projects, 2014,* PRRB Publishing, ISBN-1520207166.

Stakem, Patrick H. *Computer Architecture & Programming of the Intel x86 Family, 2013,* PRRB Publishing, ISBN-

520263724.

Stakem, Patrick H. *The Hardware and Software Architecture of the Transputer*, 2011, PRRB Publishing, ISBN-152020681X.

Stakem, Patrick H. *Mainframes, Computing on Big Iron*, 2015, PRRB Publishing, ISBN- 1520216459.

Stakem, Patrick H. *Spacecraft Control Centers*, 2015, PRRB Publishing, ISBN-1520200617.

Stakem, Patrick H. *Embedded in Space,* 2015, PRRB Publishing, ISBN-1520215916.

Stakem, Patrick H. *A Practitioner's Guide to RISC Microprocessor Architecture*, Wiley-Interscience, 1996, ISBN-0471130184.

Stakem, Patrick H. *Cubesat Engineering*, PRRB Publishing, 2017, ISBN-1520754019.

Stakem, Patrick H. *Cubesat Operations*, PRRB Publishing, 2017, ISBN-152076717X.

Stakem, Patrick H. *Interplanetary Cubesats*, PRRB Publishing, 2017, ISBN-1520766173 .

Stakem, Patrick H. *Cubesat Constellations, Clusters, and Swarms, Stakem,* PRRB Publishing, 2017, ISBN-1520767544.

Stakem, Patrick H. *Graphics Processing Units, an overview*, 2017, PRRB Publishing, ISBN-1520879695.

Stakem, Patrick H. *Intel Embedded and the Arduino-101, 2017,*

PRRB Publishing, ISBN-1520879296.

Stakem, Patrick H. *Orbital Debris, the problem and the mitigation*, 2018, PRRB Publishing, ISBN-*1980466483*.

Stakem, Patrick H. *Manufacturing in Space*, 2018, PRRB Publishing, ISBN-1977076041.

Stakem, Patrick H. *NASA's Ships and Planes*, 2018, PRRB Publishing, ISBN-1977076823.

Stakem, Patrick H. *Space Tourism*, 2018, PRRB Publishing, ISBN-1977073506.

Stakem, Patrick H. *STEM – Data Storage and Communications*, 2018, PRRB Publishing, ISBN-1977073115.

Stakem, Patrick H. *In-Space Robotic Repair and Servicing*, 2018, PRRB Publishing, ISBN-1980478236.

Stakem, Patrick H. *Introducing Weather in the pre-K to 12 Curricula, A Resource Guide for Educators*, 2017, PRRB Publishing, ISBN-1980638241.

Stakem, Patrick H. *Introducing Astronomy in the pre-K to 12 Curricula, A Resource Guide for Educators*, 2017, PRRB Publishing, ISBN-198104065X.
Also available in a Brazilian Portuguese edition, ISBN 1983106127.

Stakem, Patrick H. *Deep Space Gateways, the Moon and Beyond*, 2017, PRRB Publishing, ISBN-1973465701.

Stakem, Patrick H. *Exploration of the Gas Giants, Space*

Missions to Jupiter, Saturn, Uranus, and Neptune, PRRB Publishing, 2018, ISBN-9781717814500.

Stakem, Patrick H. *Crewed Spacecraft*, 2017, PRRB Publishing, ISBN-1549992406.

Stakem, Patrick H. *Rocketplanes to Space*, 2017, PRRB Publishing, ISBN-1549992589.

Stakem, Patrick H. *Crewed Space Stations,* 2017, PRRB Publishing, ISBN-1549992228.

Stakem, Patrick H. *Enviro-bots for STEM: Using Robotics in the pre-K to 12 Curricula, A Resource Guide for Educators,* 2017, PRRB Publishing, ISBN-1549656619.

Stakem, Patrick H. *STEM-Sat, Using Cubesats in the pre-K to 12 Curricula, A Resource Guide for Educators*, 2017, ISBN-1549656376.

Stakem, Patrick H. *Lunar Orbital Platform-Gateway,* 2018, PRRB Publishing, ISBN-1980498628.

Stakem, Patrick H. *Embedded GPU's*, 2018, PRRB Publishing, ISBN- 1980476497.

Stakem, Patrick H. *Mobile Cloud Robotics*, 2018, PRRB Publishing, ISBN- 1980488088.

Stakem, Patrick H. *Extreme Environment Embedded Systems,* 2017, PRRB Publishing, ISBN-1520215967.

Stakem, Patrick H. *What's the Worst, Volume-2*, 2018, ISBN-1981005579.

Stakem, Patrick H., *Spaceports*, 2018, ISBN-1981022287.

Stakem, Patrick H., *Space Launch Vehicles*, 2018, ISBN-1983071773.

Stakem, Patrick H. *Mars*, 2018, ISBN-1983116902.

Stakem, Patrick H. *X-86, 40th Anniversary ed*, 2018, ISBN-1983189405.

Stakem, Patrick H. *Lunar Orbital Platform-Gateway*, 2018, PRRB Publishing, ISBN-1980498628.

Stakem, Patrick H. *Space Weather*, 2018, ISBN-1723904023.

Stakem, Patrick H. *STEM-Engineering Process*, 2017, ISBN-1983196517.

Stakem, Patrick H. *Space Telescopes,* 2018, PRRB Publishing, ISBN-1728728568.

Stakem, Patrick H. *Exoplanets*, 2018, PRRB Publishing, ISBN-9781731385055.

Stakem, Patrick H. *Planetary Defense*, 2018, PRRB Publishing, ISBN-9781731001207.

Patrick H. Stakem *Exploration of the Asteroid Belt*, 2018, PRRB Publishing, ISBN-1731049846.

Patrick H. Stakem *Terraforming*, 2018, PRRB Publishing, ISBN-1790308100.

Patrick H. Stakem, *Martian Railroad,* 2019, PRRB Publishing, ISBN-1794488243.

Patrick H. Stakem, *Exoplanets,* 2019, PRRB Publishing, ISBN-1731385056.

Patrick H. Stakem, *Exploiting the Moon,* 2019, PRRB Publishing, ISBN-1091057850.

Patrick H. Stakem, *RISC-V, an Open Source Solution for Space Flight Computers,* 2019, PRRB Publishing, ISBN-1796434388.

Patrick H. Stakem, *Arm in Space*, 2019, PRRB Publishing, ISBN-9781099789137.

Patrick H. Stakem, *Extraterrestrial Life*, 2019, PRRB Publishing, ISBN-978-1072072188.

Patrick H. Stakem, *Space Command*, 2019, PRRB Publishing, ISBN-978-1693005398.

CubeRovers, A Synergy of Technologys, 2020, PRRB Publishing, ISBN-979-8651773138.

Robotic Exploration of the Icy moons of the Gas Giants. 2020, PRRB Publishing, ISBN- 979-8621431006

Hacking Cubesats, 2020, PRRB Publishing, ISBN-979-8623458964.

History & Future of Cubesats, PRRB Publishing, ISBN-979-8649179386.

Hacking Cubesats, Cybersecurity in Space, 2020, PRRB

Publishing, ISBN-979-8623458964.

Powerships, Powerbarges, Floating Wind Farms: electricity when and where you need it, 2021, PRRB Publishing, ISBN-979-8716199477.

Hospital Ships, Trains, and Aircraft, 2020, PRRB Publishing, ISBN-979-8642944349.

<u>2020/2021 Releases</u>

CubeRovers, a Synergy of Technologys, 2020, ISBN-979-8651773138

Exploration of Lunar & Martian Lava Tubes by Cube-X, ISBN-979-8621435325.

Robotic Exploration of the Icy moons of the Gas Giants, ISBN-979-8621431006.

History & Future of Cubesats, ISBN-978-1986536356.

Robotic Exploration of the Icy Moons of the Ice Giants, by Swarms of Cubesats, ISBN-979-8621431006.

Swarm Robotics, ISBN-979-8534505948.

Introduction to Electric Power Systems, ISBN-979-8519208727.

Centros de Control: Operaciones en Satélites del Estándar CubeSat (Spanish Edition), 2021, ISBN-979-8510113068.

Exploration of Venus, 2022, ISBN-979-8484416110.

Patrick H. Stakem, *The Search for Extraterrestrial Life,* 2019, PRRB Publishing, ISBN-1072072181.

The Artemis Missions, Return to the Moon, and on to Mars, 2021, ISBN-979-8490532361.

James Webb Space Telescope. A New Era in Astronomy, 2021, ISBN-979-8773857969.

www.ingramcontent.com/pod-product-compliance
Lightning Source LLC
Chambersburg PA
CBHW030449220526
45464CB00006B/2456